中国丝绸博物馆
China National Silk Museum

 编

浙江大学出版社

目录 / Contents

馆长寄语	From the Director	01
国丝馆简史	NSM's History	04
大事年表	Chronicle	06
一　丝路馆	Silk Road Gallery	10
二　非遗馆	Sericulture and Weaving Gallery	32
三　修复馆	Textile Conservation Gallery	34
四　时装馆	Fashion Gallery	36
五　女红传习馆	Textile Training Center	56
六　锦绣廊	Brocade Cafe and Museum Shop	58
七　藏品楼	Collection Building	60
八　园林	Garden	64

馆长寄语

亲爱的朋友：

欢迎来到位于杭州的中国丝绸博物馆！

从 1992 年正式开馆到现在，中国丝绸博物馆已跻身中国一级博物馆的行列。尤其是经过 2015 年起为期一年的改扩建，又于 2016 年 9 月再次对外开放。改扩建后的博物馆占地总面积达 42286 平方米，其中展陈面积约 9,000 平方米。

中国丝绸博物馆的第一个展馆是丝路馆。从湖州钱山漾遗址中出土迄今所存最早的丝绸文物，到成都老官山汉墓中出土的最早提花机，展现的是中国丝绸文化的历史演变。而丝绸之路沿途出土的各种精美丝织品，从 3 世纪至 10 世纪中国西北地区的锦绫刺绣，到 18 世纪至 19 世纪出口欧洲的外销手绘和刺绣，相信更能吸引您的目光。

若想进一步了解丝绸的制作工艺，您可以进入蚕桑馆和织造馆参观从种桑养蚕到染织刺绣的全过程。名为"中国蚕桑丝织技艺"的项目，已列入联合国教科文组织《人类非物质文化遗产代表作名录》。在这里，大家或许还有机会看到幕后的纺织品文物修复场景。我馆拥有先进的纺织品保护修复团队，来自全国的纺织品文物在这里得到保护研究，包括与大英博物馆、美国史密森学院在内的世界一流研究机构的合作项目也在这里展开。

如果对时尚感兴趣，那么一定要来看看时装馆。这里是无冕的中国时装博物馆，其中的中国时装馆展示了当代中国最顶尖的时装艺术，而西方时装馆展示的是近四百年西方时尚的变迁。临展厅中展出的是我们精心组织的专题展览，在 2016 年重新开馆之际，"锦绣世界：国际丝绸艺术展"正在这里展出，全面展示丝绸之路沿线正在使用的丝绸艺术，展现了一个锦绣世界。

中国丝绸博物馆是一个专题性博物馆，专注于国内外丝绸、纺织品、服装时尚的收藏与展示；她还是一个研究型博物馆，对本馆藏品以及世界各地的考古出土或传世保存的纺织品文物进行科学研究与修复；她也是一个活态博物馆，在这里可以看到古代织机的操作、天然染色的过程；同时，她还是一个非常国际化的博物馆，中国丝绸博物馆与世界顶尖纺织品研究机构均有合作，发起并建立了"国际丝路之绸研究联盟"，初步构建了"一带一路"沿线国家与地区关于丝绸和纺织文化的国际合作专业网络。

如果看久了需要休息，那么您可以到我们的锦绣廊中品尝咖啡，阅览图书，购置与丝绸相关的文创产品。想要购物，可以在我们的经纶堂丝绸精品店或丝博商场中选购属于自己的纪念品。

在此，我向全世界的朋友发出邀请，欢迎与我们进行学术合作，也欢迎我馆参观交流。希望中国丝绸博物馆能为丝绸文化遗产的保护、传承与创新做出贡献！

赵丰　博士 / 研究员
中国丝绸博物馆馆长

From the Director

Dear friends:

Welcome to visit the China National Silk Museum in Hangzhou.

Our museum was opened first in February 1992 and reopened in September 2016. Now it has become one of the first state-level museums in China, where you'll find 9,000 square meters of displays, divided into several galleries, in a typical southern Chinese garden of 42,286 square meters near the West Lake, a World Heritage on the UNESCO's List.

First, Silk Road Gallery will give you a journey to the Chinese silk historically and the Silk Road geometrically. Both the earliest preserved silk from Qianshanyang site, Huzhou (c.2200 BCE) and the earliest pattern loom model from Laoguanshan, Chengdu (c.100 BCE) are on display. More silk textiles are from the Silk Road, not only polychrome woven silks from the northwest of China, mainly from the 3rd to the 10th centuries, but also painted and embroidered export silks to Europe, mostly from the 18th to the 19th centuries, will attract your eyes.

For more information on how the silk is made, you can head to the Sericulture and Weaving Galleries for the whole process from mulberry planting, silkworm raising, silk releasing from cocoon, dyeing, weaving and embroidery, and the "Sericulture and Silk Craftsmanship of China" has been inscribed on the UNESCO's Representative List of the Intangible Cultural Heritage of Humanity. If you are lucky enough, you can probably visit the Textile Conservation Gallery where the silk treasures from all the country are being treated, repaired and installed or packed, since this museum owns the highest technology of textile conservation in China, and runs a lot of international joint projects with many world famous museums and institutions, such as the British Museum, and the Smithsonian, etc.

Lastly, the Fashion Gallery is the only one gallery in China focusing on the contemporary costumes. There are several sections, one for Chinese fashion of the 20th century, one for Western fashion of the recent 400 years, and a temporary special exhibition, A World of Silks, of all the silks from the world on the occasion of the reopening.

So the China National Silk Museum is a specialized museum, focusing on the collection and exhibition of silk, textile, costume and fashion, not only Chinese, but also abroad. It is a research museum, doing laboratory research and conservation work, not only for museum collection but also new findings from archaeological field. It is a living museum, weaving on traditional looms, dyeing with natural materials and creating silk products with new designs. It is an international museum which has sent several silk exhibitions touring around the world and jointly founded the International Association for the Study of Silk Road Textiles, including around 30 institutes from 15 countries.

To rest your feet, you can go to the Brocade Café located in the central area of the museum. It features drinks, books, souvenirs and silk products such as clothing, decorations, and home silk-making gear. For shopping, you can also visit the Jingluntang Silk Boutique located in the basement of Fashion Gallery for high-quality design silk products, and Sibo Silk Shop for local silk products located in the ground floor of the Silk Road Gallery.

We are waiting for more collaboration with more friends from the world, not only the academic side but also common tourists. We hope the China National Silk Museum could do more on the preservation, transmission and innovation of the silk cultural heritage in the future.

Dr/Prof. Feng ZHAO

Director of China National Silk Museum

国丝馆简史

　　位于杭州西子湖畔玉皇山下的中国丝绸博物馆（国丝馆）是国家一级博物馆，中国最大的纺织服装类专业博物馆，也是全世界最大的丝绸专业博物馆。现占地面积42,286平方米，建筑面积22,999平方米。中国丝绸博物馆于1992年2月26日建成开放，2004年1月1日起对公众实行免费开放。2015年又开启了改扩建工程，经过一年的风雨兼程，2016年9月国丝馆以全新的面貌呈现给国内外参观者。

　　经过几代丝博人的共同努力，国丝馆在征集丝绸藏品、举办国内外展览、保护纺织品文物、传承蚕桑丝织技艺、开展丝绸科普教育、弘扬丝绸文化等方面取得了令人瞩目的成绩。人类非物质文化遗产代表作"中国蚕桑丝织技艺"由此申报，"纺织品文物保护国家文物局重点科研基地"落户此地。近年来，国丝馆与世界各地的学术机构加强合作，成立了"国际丝路之绸研究联盟"，开展了大量的合作项目，正在让精美的丝绸和博大的丝绸文化，沿着丝绸之路，走向世界，走向人类的美好明天！

NSM's History

China National Silk Museum (NSM), near the West Lake in Hangzhou, is one of the first state-level museums in China , covering an area of 42,286 square meters and a building area of 22,999 square meters. It was opened on February 26th, 1992, and welcomes visitors free of charge since January 1st, 2004. In 2015, the museum started an extension and renovation project, after a year of trials and hardships, NSM finally embraces the audience from all over the world in a brand new look.

Through the joint effort of all the staff, China National Silk Museum achieved remarkable progress in obtaining silk collections, hosting domestic and foreign exhibitions, protecting textile heritages, inheriting sericulture and silk weaving skills,carrying out popular silk science education and promoting the silk culture. In the recent years, the museum has cooperated with academic institutions around the world, and jointly founded the International Association for the Study of Silk Road Textiles, which has undertaken a large number of international cooperation, and is bringing the exquisite silk and the broad silk culture to all over the world.

中国丝绸博物馆原大门   The original gate of China National Silk Museum

大事年表

中国丝绸博物馆建设项目立项

Project approved by the government

1986

建成开放

Opened to the public

26 February
1992

开始对公众实行免费开放

Admission-free to visitors

2004

1987

2000

开工奠基

Construction started

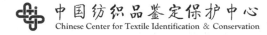

中国纺织品鉴定保护中心
Chinese Center for Textile Identification & Conservation

纺织品文物保护国家文物局重点科研基地
Key Scientific Research Base of Textile Conservation, SACH

2012

"国际丝路之绸研究联盟"成立

IASSRT established

October
2015

2009

2015
August

2016
September

全面闭馆，启动改扩建工程
Closed for renovation

新馆启用
New museum opened to the public

一　丝路馆　Silk Road Gallery

丝路馆以"锦程：中国丝绸与丝绸之路"为线索，讲述中国丝绸走过的五千年光辉历程及其丝绸从东方传播至西方的万里丝路。

Silk Road Gallery will give you a journey to the Chinese silk historically and the Silk Road geometrically.

▼《印象丝路》 挂毯
施慧
中国美术学院万曼壁挂研究所

Silk Road Impression Tapestry
by Shi Hui
Research Institute of Art Tapestry Varbanov, China Academy of Art

▲ 大厅　Main Hall

锦程：中国丝绸与丝绸之路

中国是世界丝绸的发源地，以发明植桑养蚕、缫丝织绸技术而闻名于世，被称为"丝国"。数千年来，中国丝绸以其独有的魅力、绚丽的色彩、浓郁的文化内涵，为中国文明谱写了灿烂篇章。同时，丝绸催生了丝绸之路。作为丝绸之路的主角，丝绸产品及其生产技术和艺术成为丝绸之路上最重要的内容，被传播到了世界各地，为东西方文明互鉴做出了卓越的贡献。从史前走来的中国丝绸，与中华文明相伴相生，直至今日依然绚烂如花。

丝路馆是中国丝绸博物馆的主要展厅，其中分为序厅、二楼展厅和三楼展厅，二楼包括源起东方（史前时期）、周律汉韵（战国秦汉时期）、丝路大转折（魏晋南北朝时期）、兼容并蓄（隋唐五代时期）、南北异风（宋元辽金时期）五个单元，三楼包括礼制煌煌（明清时期）、继往开来（近代）、时代新篇（当代）三个单元。展览以"锦程——中国丝绸与丝绸之路"为题，通过丝绸之路沿途出土的汉唐织物等精品文物300余件，讲述了中国丝绸走过的五千年光辉历程及丝绸从遥远的东方传播至西方的万里丝路，展示中国丝绸对人类文明的贡献。

The Way of Chinese Silk: Silk History and the Silk Road

China, the home of silk, was known to the ancient Greeks and Romans as "Serica," the "Land of Silk". Sericulture, or silk making, including the cultivation of mulberry trees and silkworms, and the techniques of reeling, spinning, dyeing and weaving silk fibers, were practiced in China thousands of years ago. Silk has long played a major role in the Chinese civilization and was a key factor in the creation of the Silk Road, which linked the civilizations of East and West. Chinese silk is almost as old as the Chinese civilization itself, and has evolved hand in hand with the Chinese people and the Chinese nation. Today Chinese silk continues to reflect its past glory. The exhibition is on the clue of "The Way of Chinese Silk: Silk History and the Silk Road", divided into eight sections, using the Han and Tang dynasties' textiles excavated along the Silk Road to tell the story of 5,000 years of glorious history which Chinese silk traveled and how it transmitted from the Far East to the West.

The exhibition on the second floor has 5 sections which are Early Origins (Neolithic period); Sericulture in the Warring States period, Qin and Han dynasties (BCE475); Silk in the Wei, Jin and Northern and Southern dynasties (220–581); Sui, Tang and Five Dynasties period (581–960); Silk in the Song, Yuan, Liao and Jin dynasties (907–1368), illustrating history of Chinese silk from prehistory to the Song and Yuan dynasties and the cultural exchanges along the Silk Road during different time periods.

The third floor has three sections which are Silk in the Ming and Qing dynasties (1368–1911); Chinese Silk in the Modern Age (1911–1949) and Silk in Contemporary China (1949 to the present), presenting the development of silk from the Ming and Qing dynasties to the present age. For Ming and Qing dynasties, displayed here are the high-grade fabric of *zhangduan* (patterned) velvet and *zhangrong* (voided) velvet; for feudal etiquette displayed here are dragon robe, *mang* (four–claw) dragon robe, robe material, rank badges and official uniforms of the Ming and Qing dynasties and the export of silk embroidery of late Qing Dynasty. Each time period has its representative works.

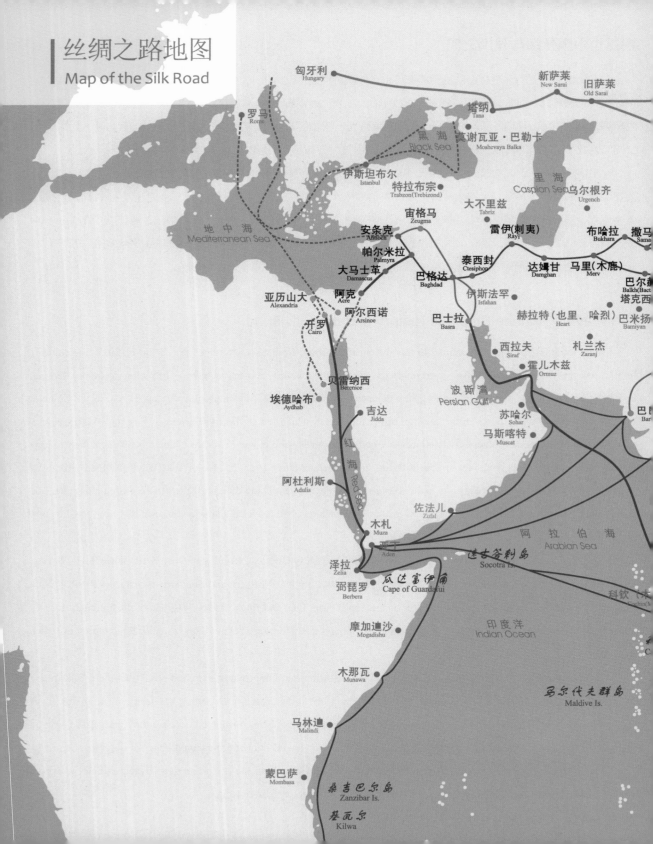

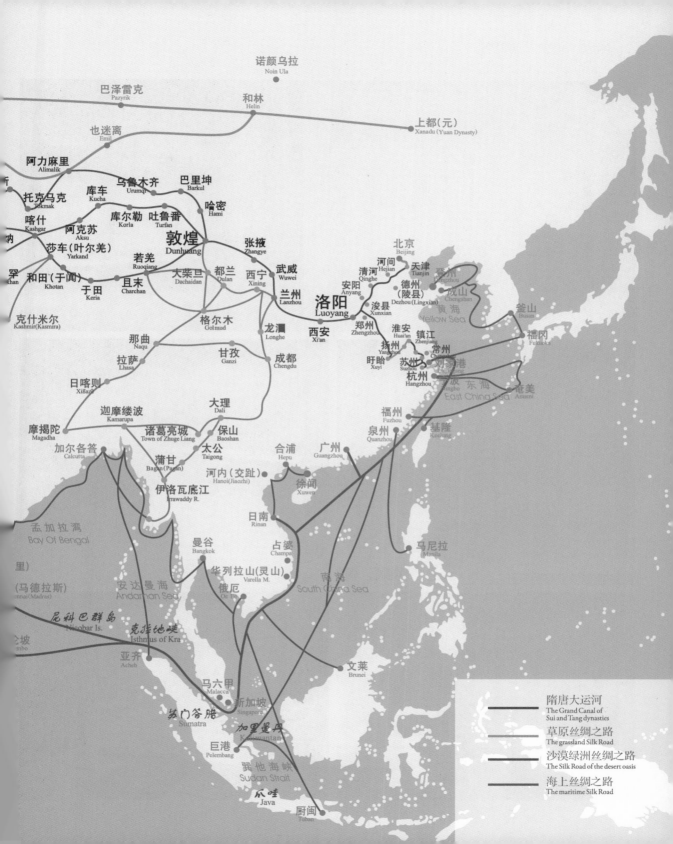

二楼展厅 Second Floor Gallery

▲ 几何纹锦　　战国
Brocade with geometric patterns
Warring States period (BCE475—BCE221)

▲ "恩泽"锦　　汉晋
Brocade inscribed "Mercy and plentitude"
Han or Jin Dynasty (BCE206—ACE220)

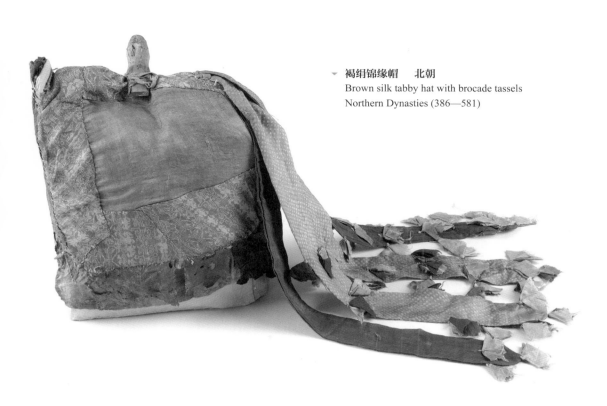

▼ 褐绢锦缘帽　北朝
Brown silk tabby hat with brocade tassels
Northern Dynasties (386—581)

▲ "长葆子孙"锦缘绢衣裤　汉代
Jacket and pants set, silk tabby and brocade
Han Dynasty (BCE206—ACE220)

▲ 立狮宝花纹锦　唐代
Samite with standing lion and precious flower
Tang Dynasty (618—907)

▲ 暗红地联珠新月纹锦覆面　北朝
Burial face cover, brocade with linked pearl and crescent　Northern Dynasties (386—581)

▲ 紫褐色罗印金彩绘花边单衣　南宋
Unlined brown gauze robe with borders of gilded and painted motifs
Southern Song Dynasty (1127—1279)

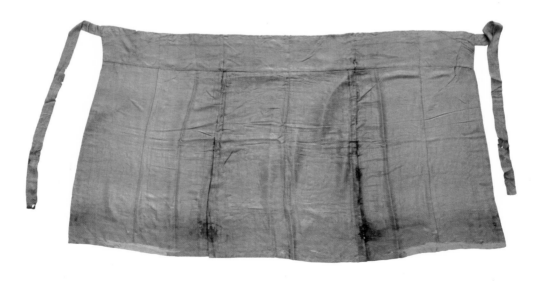

▲ 折枝朵花纹罗裙　南宋
Gauze skirt, with branch and flower motifs
Southern Song Dynasty (1127—1279)

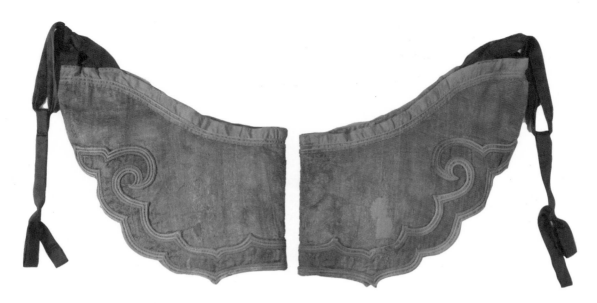

▲ 纳石失靴套　元代
Nasij boot covers
Yuan Dynasty (1271—1368)

▲ 缎地刺绣花卉纹枕顶　元代
Embroidered satin pillow panels with floral motifs
Yuan Dynasty (1271—1368)

三楼展厅 Third Floor Gallery

▲ 明代钱氏墓出土女服　明代
　Women's wear
　Ming Dynasty (1368—1644)

▶ 缂丝孔雀纹圆补　明代
　Kesi tapestry round rank badge with peacock motif
　Ming Dynasty (1368—1644)

▲ 诰命　清代
　Imperial edict
　Qing Dynasty (1644—1911)

▶ 红地刺绣花卉旗服　清代
　Manchu robe with embroidered floral motifs
　Qing Dynasty (1644—1911)

▲ 明黄色团龙纹实地纱盘金绣小龙袍　清代
Embroidered dragon robe for imperial prince
Qing Dynasty (1644—1911)

▶ 紫地缂丝龙袍　清代
Dragon robe, *kesi* tapestry
Qing Dynasty (1644—1911)

▸ **大红绸地盘金彩绣八团龙女袍　清代**
Woman's robe embroidered with eight dragon medallions
Qing Dynasty (1644—1911)

▸ **黑缎彩绣对枝桃花纹褂襕　清代**
Dress embroidered with peach blossom motif
Qing Dynasty (1644—1911)

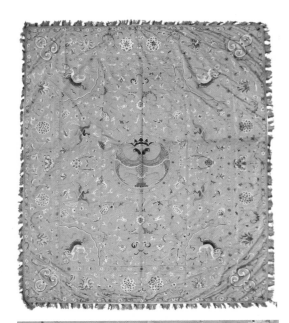

▲ 绿地妆花蟒缎袍料　清代
Brocaded robe material with *mang* (four-claw) dragon motif
Qing Dynasty (1644—1911)

▲ 外销黄缎地彩绣双头鹰花鸟纹床罩　18世纪晚期
Bed cover, embroidered with double-head eagle, and floral motifs on yellow satin , for export　Late 18th century

外销白缎地彩绣人物伞 19世纪60—70年代
Parasol embroidered with human figures in garden, for export 1860s—1870s

▲ 《西湖》像景　20 世纪早期
　 West Lake, Woven silk pictures　Early 20th century

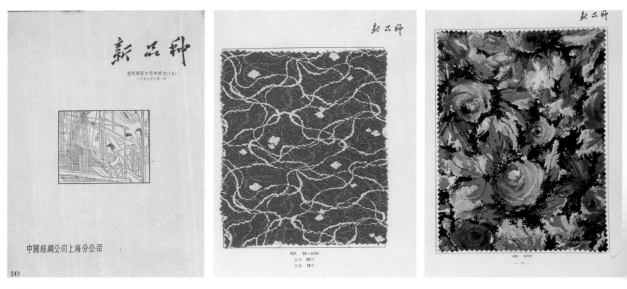

▲ 《新品种》样册　1959—1960
　 Silks Sample books　1959—1960

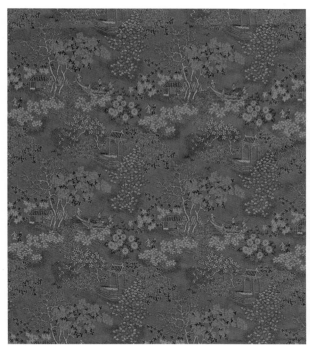

▲ "江南水乡"古香缎被面　20世纪70年代
　 "South of the Yangtze River", Quilt cover　1970s

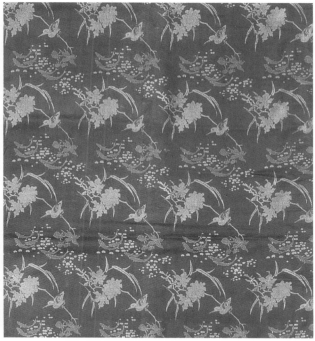

▲ 紫地金鱼花卉纹缎匹料　民国
　 Satin damask with goldfish and floral motifs on purple ground, bolt silk
　 The Republic of China（1911–1949）

▲《变》数码织锦　2013年
　 Change, Digital woven　2013

二 非遗馆　Sericulture and Weaving Galleries

非遗馆包括蚕桑馆和织造馆两个空间，总名"天蚕灵机"，以列入联合国教科文组织人类非物质文化遗产代表作名录的"中国蚕桑丝织技艺"为中心，分"天虫作茧""蚕乡遗风""制丝剥绵""染缬绘绣"和"天工机织"五大部分娓娓展开，并选取270余件展品，展示了中国蚕桑丝织技艺非物质文化遗产涵盖的蚕桑、习俗、制丝、丝织、印染、刺绣技艺的方方面面，是博物馆全面系统保护、传承和展示人类非物质文化遗产的一次尝试。

Sericulture and silk technology originated in China, and serve as a cultural symbol of the Chinese nation. "Sericulture and Silk Craftsmanship in China" has been inscribed on the UNESCO's Representative List of the Intangible Cultural Heritage of Humanity on September 28, 2009. The exhibition displays all the aspects about mulberry cultivation, silkworm breeding and silk reeling, dyeing and weaving. There are five sections of the exhibition, which are "The Story of the Silkworm", "Folk Customs in the Birthplace of Sericulture", "Silk-making Techniques", "Textile Printing, Dyeing and Embroidery" and "Weaving Techniques", displaying more than 270 objects. The exhibition technique combines static display with live demonstration, helping the audience better understand the intangible cultural heritage.

▼ 蚕桑馆　Sericulture Gallery

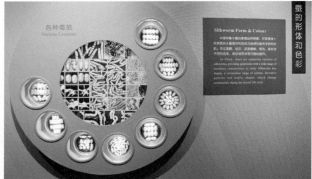

织造馆　Weaving Gallery

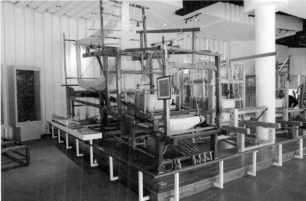

三 修复馆　Textile Conservation Gallery

　　中国丝绸博物馆是中国古代纺织品鉴定保护中心和纺织品文物保护国家文物局重点科研基地的所在地。为了进一步门普及纺织品文物科技的知识，将幕后的修复工作展示出来，中国丝绸博物馆特地设立了纺织品文物修复展示馆，这也是国家文化遗产保护科技区域创新联盟（浙江省）的示范应用基地。修复馆的一楼用于纺织品的信息提取、修复、研究等工作，观众可从二楼观看专业人员正在开展的文物保护修复的全过程。同时，二楼上还不定期地举办纺织品文物的保护研究成果展示。

　　China National Silk Museum is where the Chinese Center for Textile Identification and Conservation is located, as well as the Key Scientific Research Base of Textile Conservation, State Administration of Cultural Heritage. To present the audience how the conservation of cultural heritage takes place, the museum set up a Conservation Gallery. The first floor is for the lab and the audience can watch the whole process from the second floor. The second floor is also used for exhibition on the subject of the protection, research and restoration of textiles and cultural heritage.

1 丝府宋韵：黄岩南宋赵伯澐墓出土服饰展
　The Sound of Silk: Costumes found in the Tomb of Zhao Boyun (1155—1216)

2 旧旗新帜：上海市历史博物馆藏纺织品文物保护修复展
　Old Clothing and New Flag: Textiles and Costumes from the Collection of Shanghai History Museum

3 绽放：蕾丝的前世今生
　Blooming: The Past and the Present of Lace

4 千缕百衲：敦煌莫高窟出土纺织品的保护与研究
　Thousands of Thread and Hundreds of Patchwork: Conservation and Research on Excavated Textiles from Mogao Grottoes at Dunhuang

5 钱家衣橱：无锡七房桥明墓出土服饰保护修复展
　Qian Family's Wardrobe: Costume Found in the Tomb of Qian Zhang (1486—1505) and his wife

6 丝路之绸：新疆纺织品文物修复成果展
　Conservation of Textiles from the Silk Road in Xinjiang

四 时装馆　Fashion Gallery

▲ 天罗　刘君
Celestial Net　by Liu Jun

▲ 化蝶　周力
Rebirth　by Zhou Li

更衣记：中国时装艺术

　　中国时装艺术自 20 世纪开始，此时中国的社会政治、经济、文化都发生了巨大的变化。人们推翻了中国最后一个封建王朝，从此，中国服装摒弃了森严等级制度，吸收了西方服饰文化，开启了百年服饰新时尚。

　　"更衣记"记录了百年中国服装变迁的脉络。1920 年前后，中国时装的先驱——旗袍逐渐形成和成熟，成为历久弥新的时尚经典。1950 年后，日常服装都打上了革命的烙印，蓝衣绿服是中国社会形态的缩影。改革开放之后，中国服装设计走过了快速发展的 40 年，在融入世界的浪潮中，中国出现了一批服装设计师和一批具有中国特色和国际水准的作品，拥有了知名品牌，也出现了真正的时尚产业。

An Evolution of Fashion: Chinese Costume from 1920s to 2010s

The 20th century was a significant century in which Chinese politics, economics, and culture experienced tremendous changes. People overthrew China's last feudal dynasty and from then on, Chinese costumes have ceased to reflect the rigid hierarchy system and adopted Western clothing culture. This fusion of styles helped creating a new Chinese fashion.

In this exhibition, we've tried to record the process of change in Chinese fashion.Since the 1920s, Qi-pao, recognized as the most representative of Chinese women's costumes, was gradually developed and improved,it has become a classic in China's history of clothing.1950s,outfit design was hugely influenced by the revolutionary times. Blue and green clothes became the microcosm of the Chinese society.After the reform and opening up, China's fashion design experienced 30 years of rapid development, during which a fashion industry emerged. To integrate with the global trends, China has cultivated many costume designers, who have created a large number of international standard works with Chinese characteristics, and a real fashion industry appeared.

▲ 花卉纹绸倒大袖短袄黑色小花纱裙　20世纪20年代
　Damask lined-jacket with round hem and gauze skirt　1920s

▲ 卷云纹倒大袖旗袍　20世纪20年代
　Super-sleeve qipao woven with coiled clouds　192

◄ 倒大袖旗袍马甲　20世纪20年代
　Inverted sleeve qipao vest　1920s

◂ 花卉纹蕾丝无袖旗袍　20 世纪 20 年代
Sleeveless qipao bound with floral motif lace　1930s

▴ 婚纱礼服　20 世纪 30 年代
Wedding dress　1930s

◂ 印花绸短袖旗袍　20 世纪 30 年代
Printed short-sleeve qipao　1930s

▲ 印花绸长袖旗袍　20世纪40年代
Printed long sleeve qipao　1940s

▶ 印花绸短袖旗袍
　 包陪庆女士捐赠
Printed short sleeve qipao
Donated by Ms. Bao Peiqing

▲ 海虎绒大衣和花卉纹印花旗袍　20世纪40年代
Plush wool-velvet Overcoat qipao with printed floral
motif and coat　1940s

《鼎盛时代》 吴海燕 1993年
The Golden Age by Wu Haiyan, 1993

▲ 《敦煌》
劳伦斯·许
Color of Dunhuang
by Lawrence Xu

▶ 《寻凤行凤循凤》
裘海索
Phoenix searcher, phoenix walker, phoenix follower
by Qiu Haisuo

◀ 《长江系列》之一
张肇达
The Yangtze River
by Zhang Zhaoda

▲ 翘摆凤尾裙
郭培
Phoenix tail
by Guo Pei

从田园到城市：400 年的西方时装

中国丝绸博物馆从馆藏近 4 万件西方时装中甄选了 400 余件精品，囊括了欧美 17—20 世纪各重要时期的服饰品，分五个时期展示了西方时装四百年的发展轨迹、时代特征、服饰风格以及时装与艺术的关联和影响。

展出的大部分展品为西方服饰史中的代表性服饰，或具备该时期服饰的典型特征。包括 17 世纪巴洛克礼服裙，18 世纪华托服、波兰裙、帕尼尔廓形的礼服裙以及 19 世纪帝政时期的简·奥斯丁裙、巴瑟尔裙等。20 世纪展品中有半数出自扬名史册的杰出设计师之手，如 Jeanne Lanvin、Chanel、Christian Dior、Balenciaga、Givenchy 和 Pierre Balmain 等。另外单设服饰品展示区块，展出 19 世纪末至 20 世纪精美鞋子、手包、首饰、化妆用具等。

From Rural to Urban: 400 Years of Western Fashion

Western fashion is the main stream and the core of modern fashion. Even though diverse perspectives in fashion, such as the trends from Asia, Africa and South America bring sub-trends and start to create the fascinating fashion kaleidoscope, philosophy of the Western fashion still dominates this international market. China National Silk Museum carefully selected 455 pieces out of about forty thousand collections, for display to the audience. The display contains representative clothing and the accessories from the 17th century to the 20th century, divided into five sections, showing development track, time characteristics, fashion style and links of fashion and art.

Most objects are representative clothing at that time period, including formal dress from the Baroque era, Watteau gown and Pannier from the 18th century and Empire silhouette from the 19th century. The collections from the 20th century are mostly from outstanding designers such as Jeanne Lanvin, Gabrielle Bonheur Chanel, Christian Dior, Cristóbal Balenciaga, Hubert de Givenchy, Pierre Balmain, etc. In addition to that, there is an accessory section which shows the exquisite shoes, purses, jewelries, cosmetics from the end of the 19th century to the 20th century.

▸ 礼裙　1680—1700 年
　Formal Dress　1680—1700

▴ 长礼服和马甲　法国　1790 年
　Gown and waistcoat　France　1790

▸ 裙　19 世纪 50—70 年代
　Dress　1850s—1870s

▸ 裙　皮埃尔·巴尔曼　1958年
　Dress　by Pierre Balmain　1958

▴ 裙　克里斯汀·迪奥　1954年
　Dress　by Christian Dior　1954

▸ 外套　佛朗哥·莫斯基诺　20世纪90年代
　Coat　by Franco Moschino　1990s

临展厅

位于时装馆的临展厅专用于举办临时陈列，主展厅约 600 平方米。已举办过"锦绣世界：国际丝绸艺术精品展""化蝶：2016 时尚回顾展""桂风壮韵：广西壮族织绣文化展""古道新知：丝绸之路文化遗产保护科技成果展""荣归锦上：17 世纪来的法国丝绸等展览"，内容涉及古今中外各个领域。

Temporary Exhibition Gallery

Located in the Fashion Gallery, the Temporary Exhibition Gallery is dedicated to temporary exhibitions. The main exhibition hall is about 600 square meters. It has held "A World of Silks: International Silk Art Exhibition", "Transform Into Butterfly —2016 Fashion Review", "Brocade and Embroidery Culture of Zhuang Nationality in Guangxi", "New Knowledge on Ancient Road: Silk Road Cultural Heritage Sci-tech Achievements", "Glory on Silk: French Textiles (1700 to the Present)" and other exhibitions, covering all aspects, from ancient to modern, China to foreign countries.

1—2	锦绣世界：国际丝绸艺术展 A World of Silks: International Silk Art Exhibition	
3—4	荣归锦上：1700年以来的法国丝绸展 Glory on Silk: French Textiles (1700 to the Present)	
5—6	古道新知：丝绸之路文化遗产保护与研究成果展 New Knowledge on Ancient Road: Silk Road Cultural Heritage Sci-tech Achievements	
7—8	匠·意：2017年度时尚回顾 To Craftsmanship — 2017 Fashion Review	
9	桂风壮韵：广西壮族织绣文化展 Brocade and Embroidery Culture of Zhuang Nationality in Guangxi	

新猷资料馆

新猷资料馆是以蚕桑丝绸界老前辈朱新予和蒋猷龙两位先生的名字命名的纺织信息中心，收集和展示现代纺织面料样本、纺织人物档案和蚕桑丝绸史、染织服装史、纺织考古、丝绸之路相关的有历史价值的中外报刊书籍、音像材料。这里也举行小型的文献展，如"弦歌不辍：浙江敦煌学与丝绸之路研究文献展"等，同时，这里也是中国丝绸博物馆对公众进行讲座的主要场地，平均每周都有1—2场专业的学术报告。

Xinyou Archive Center

Xinyou Archive Center is a textile information center named after Zhu Xinyu and Jiang Youlong, two important predecessors in the sericulture and silk industry. Samples of modern fabrics, precious personnel archives, historical newspapers, books and audio materials in Chinese and other languages concerning silkworm breeding and silk reeling, the history of silk, dyeing techniques, the history of garment, textile archaeology and the Silk Road are on display. The center also deals with recording, assorting and provision of pictures and information with the aid of modern technology.

纸上衣影
西方时装插画展

展览时间：2017 年 10 月 23 日—12 月 22 日
展览地点：新猷资料馆

20 世纪初，为了对抗摄影的高速度与低成本，彩色插画技术应运而生了。这项技木操作相对简单，成本更为低廉，极少量复制，时尚艺术家们开始欣赏它的这力，并通过这种富有创新精神的插画寻找新出路。成纸张变得越来越轻薄巧妙。时装插画结合十分细腻精准，这项新兴艺术给时尚界插画提出了一种全新的视角，我们可以看到在20 世纪初中国以其独特的风采，甚至可以说，这就是一部结合时尚史、艺术史与女性解放史。

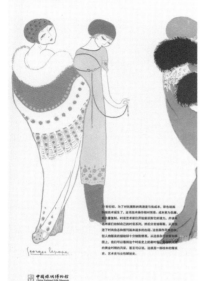

丝绸盛开迎春花
20 世纪下半叶的上海第七印绸厂

展览时间：2017 年 12 月 22 日—2018 年 03 月 05 日
展览地点：新猷资料馆

近代上海的丝绸印花业起步于 20 世纪初，后因由于战争的缘故和商品的短缺，到解放前已陷入低谷。新中国的成立为丝绸印花业带来了新的发展契机。从 50 年代的社会主义国家销售传统，设计工艺，绝色和整理等方面的革故鼎新使产品得以成长协作，到 70 年代中期时，已具备了多与国际时尚接轨的实力。本展览呈出的正是其中的一个例子——上海第七印绸厂及其身为大花印绸厂的姐妹文献，从这些产品的新闻中，我们可以展出 20 世纪下半叶中国丝绸印花业的轮廓。

竞芳
骆竞芳丝绸设计作品

展览时间：2018 年 03 月 19 日—2018 年 05 月 20 日
展览地点：新猷资料馆

这个展览纪念骆竞芳诞辰九十年之际，2009 年，我馆举办了名为《笔歌与连绵——1957-1978 年中国丝绸设计师展》的展览，展出了近二十多年间中国丝绸图案设计师的精选，并向观众们陈述了中国近来丝绸事业的成就和代表典范。骆竞芳是了这些老设计师们的其中一位，她们中又为我们留业和自己的。但是，由于国域所限，当时只具有三十份展品和以展出、今天，通过更新整理，我们从她捐赠的 130 余件设计精中挑选出两件，并希望能欲求引起您对 20 世纪下半叶中国丝绸印花业的一点探究之心。

银瀚厅

银瀚厅是可以举办国际会议、走秀、讲座、演出以及作为展示空间的多功能厅，约400平方米。已举办过"锦绣世界""古道新知"等国际会议，举办过中东欧艺术家的"丝绸与传统"展，举办过2017年杭州国际旗袍日沙龙、2018年国丝汉服节，特别是作为中国丝绸博物馆"国丝之夜"中的主打品牌"丝路之夜：丝绸之路的跨文化交流"的主场地，举办过法兰西、意大利、阿拉伯、波斯、长安、敦煌、天山、中东欧、蕾丝之夜等系列活动。

Galaxy Hall

Galaxy Hall is a multi-purpose hall that can hold international conferences, catwalks, lectures, performances, and exhibitions, covering an area of about 400 square meters. It has held international conferences such as "A World of Silks: " and "New Knowledge on Ancient Road", "Silk and Tradition: China and Central & Eastern Europe Contemporary Art Silk" exhibition, the 2016 China Fashion Award Ceremony, and the 2017 Hangzhou Global Qipao Festival Salon. It was the main venue for the activity of " NSM's Evening", and "Evening on the Silk Road: Intercultural Dialogue on the Silk Road", including Franch Evening, Italian Evening, Arabian Evening, Persian Evening, Chang'an Evening, Dunhuang Evening, Tianshan Evening, Central and Eastern European Evening, Lace Evening and other activities.

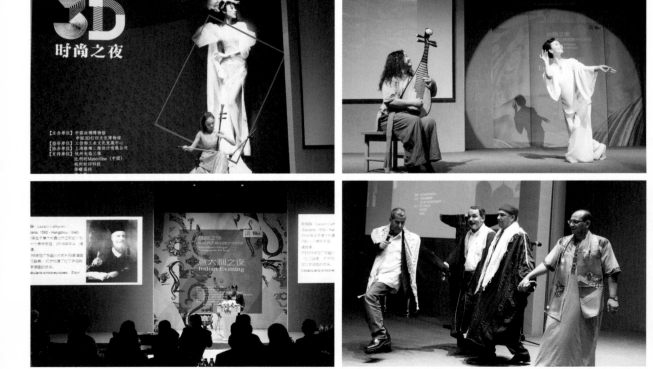

经纶堂精品店

经纶堂是中国丝绸博物馆与浙江凯喜雅集团合作成立的一家文化创意公司,专以文化遗产为主题、天然丝绸为材质,迄今已生产了几十款丝巾产品,其中一些也曾作为G20杭州峰会国礼产品。经纶堂丝绸精品店位于时装馆楼下,是展示和销售丝绸文化商品的一个集合地。

Jingluntang Boutique Shop

Jingluntang is an art company jointly established by China National Silk Museum and Zhejiang Cathaya Group. Its products are exclusively based on the theme of cultural heritage and are made of natural silk. So far, it has produced dozens of products of scarves, including the G20 Hangzhou Summit edition. The Jingluntang Boutique Shop is located downstairs in the Fashion Gallery and is a place where silk cultural products are displayed and sold.

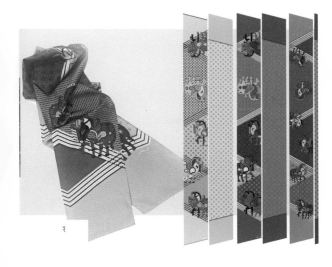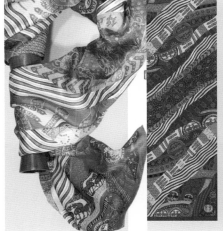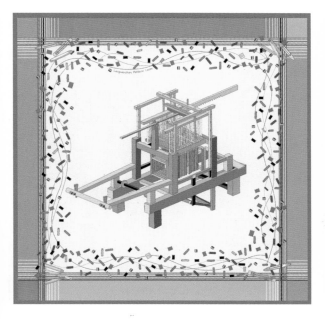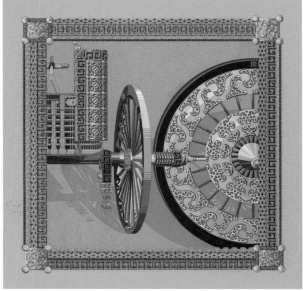

五 女红传习馆　Textile Training Center

女红指在男耕女织的古代社会中女子要从事的一些以印染、织绣、编缝为主的工作。随着机器大生产的发展，传统的女红技艺逐渐消失在人们的日常生活中。针对中国传统纺织文化传承和创新的现状，中国丝绸博物馆特别在G20杭州峰会期间设立了女红传习馆，一是培养兴趣，对少年儿童进行寓教于乐的教育，从小埋下其做女红的种子；二是普及知识，传播较为经典的相关知识，供其进行课外实践；三是较高层次的切磋技艺、研习技法和设计，把传统女红作为创新、创业的重要部分。

女红传习馆位于丝路广场楼下，主要开设织、染、绣、编等与纺织服饰相关的专业课程。传习馆的第一批小学员在G20期间为多国第一夫人表演了织造技艺。之后又举办过扎染、彩灯、蕾丝、迷你时装、植物染色、综版织等多场传习课程，受到极大欢迎。目前，女红传习馆邀请中国著名染织美术教育家、中国丝绸博物馆理事会名誉理事长常沙娜老师题写了馆名，又在杭州大关小学设立了校园基地，在海宁云龙村设置了乡村实践基地。

Textilework refers to dyeing, weaving, and embroidery that were mainly engaged in by women in the ancient societ when the men ploughed while the women wove. With the development of machine production, traditional hand weaving process is stepping down from the stage of the history and disappearing in people's daily life gradually. Based on the present situation of the inheritance and innovation of Chinese traditional textile culture, and oriented toward the broad audience, China National Silk Museum has set up the Textile Training Center which offers the professional courses related to weaving, dyeing, embroidering and braiding after the completion of the new museum so as to satisfy the demand of the people who love the traditional textile culture and to inherit the traditional skills.

六 锦绣廊　Brocade Cafe and Museum Shop

　　锦绣廊是中国丝绸博物馆最为重要的观众服务空间,位于修复馆和时装馆之间、锦绣广场之侧,其中包括锦廊咖啡、锦道文道和晓风书屋等多项内容,不仅是展示和销售文创产品和相关书籍的重要场所,也是观众得以憩歇和休闲的极佳去处。

　　Besides the Brocade Cafe which offers the visitors a place to take a rest, there are also some rest places of various sizes and a bookstore and a souvenir shop inside the museum. With the exquisite landscapes, China Nation Silk Museum has become a perfect place for pleasure.

七 藏品楼　Collection Building

纺织品文物保护国家文物局重点科研基地

2010年10月，国家文物局批准以中国丝绸博物馆为依托单位设立纺织品文物保护国家文物局重点科研基地，研究方向为纺织品文物保护，研究内容包括纺织品相关文物分析检测鉴定、保护修复关键技术研究、传统工艺与价值挖掘等。同时，基地在浙江理工大学设立联合实验室，还在新疆（2011年）、甘肃（2015年）、内蒙古（2017年）设立了工作站，在西藏（2014年）设立了联合工作站。历年来，基地承接了大量丝绸之路沿线的纺织品考古、保护、研究项目，如敦煌丝绸艺术全集、敦煌莫高窟出土丝绸、新疆营盘墓地出土服饰、乌兹别克斯坦蒙恰特佩出土丝绸、俄罗斯北高加索地区合作纺织考古等，并于2017年发起倡议成立了国际丝路之绸研究联盟。

Key Scientific Research Base of Textile Conservation, SACH

In October 2010, the State Administration of Cultural Heritage approved the establishment of the Key Scientific Research Base of Textile Conservation by the China National Silk Museum. The research direction is the protection of textile heritage. The research includes the analysis, testing and identification of textile-related cultural heritage, key technologies for protection and conservation, and the discovery of traditional craftsmanship and its value. At the same time, the base has established a joint laboratory at Zhejiang Sci-Tech University and established workstations in Xinjiang (2011), Gansu (2015), Inner Mongolia (2017), and a joint workstation in Tibet (2014). Over the years, the base has undertaken a large number of textile archaeology, conservation, and research projects along the Silk Road, such as the complete collection of The Textiles from Dunhuang, the silk unearthed from the Mogao Grottoes in Dunhuang, the costumes unearthed from the Yingpan tomb in Xinjiang, the silk unearthed in Uzbekistan, and the cooperation of the archaeological textiles in the northern Caucasus in Russia. In 2017, it initiated the founded Alliance on Technological Innovations of Cultural Heritage along Silk Road (ATICS).

实验室

实验室已配备激光共聚焦显微镜、液相色谱质谱联用仪、分光测色仪、扫描电镜－能谱、氨基酸分析仪、红外光谱显微镜、氙灯试验箱等大型精密成套仪器设备，基本满足纺织品文物形貌、结构、成分等相关研究需求。利用自有大型仪器设备，为国内外机构提供基于形貌和 FTIR 的纤维鉴别、基于 UFLC–PDA–MS 的染料测试、基于 SEM–EDS 的颜料元素分析、基于 HPLC 的氨基酸测试等多种技术服务，先后出具近百份测试报告。

实验室从纤维和染料等层面对纺织品文物开展深入研究，不仅能够揭示纺织品文物的重要科学内涵，提升纺织品文物的认知水平，还拓展了纺织品保护研究的深度和广度。

Laboratory

The laboratory is equipped with laser confocal microscope, high performance liquid chromatography coupled with mass spectrometer, spectrophotometer, scanning electron microscope-energy dispersive X-ray spectrum, amino acid analyzer, infrared microscope, xenon test chamber and other large-scale precision instruments, which basically meet the relevant research needs of textile cultural relics such as morphology, structure, and composition. Using its own large-scale precision instruments, the laboratory has provided various technical services for domestic and foreign organizations such as fiber identification based on morphology and FTIR, dye identification based on UFLC-PDA-MS, pigment analysis based on SEM-EDS, amino acid testing based on HPLC, and has issued nearly one hundred test reports.

The in-depth research of the laboratory on textile cultural relics, focusing on the fiber and dyestuff, not only reveals the important scientific connotations of textile relics, improves the cognition of textile relics, but also expands the depth and breadth of textile protection research.

 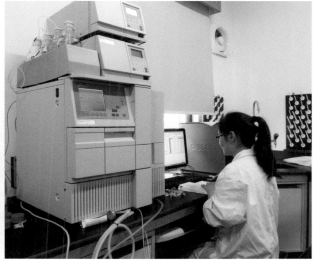

修复室

修复室目前配备有特制修复工作台、无频闪修复灯、三维视频显微镜、高保真信息采集系统、低压清洗台、染色机等设备，可以承担保护修复全过程的技术要求，包括信息采集、分析检测、现状评估、清洗染色、平整修复、包装保存等环节，目前已为全国 50 家文博机构提供保护修复技术服务。

Conservation Studio

The conservation studio is currently equipped with special repair workbenches, no stroboscopic repair lights, three-dimensional video microscopes, high-fidelity information acquisition systems, low-pressure cleaning stations, dyeing machines, and other equipments, and can undertake the technical requirements for the entire process of conservation and restoration, including analysis and testing, status assessment, cleaning and dyeing, leveling and repairing, and packing preservation. It has provided protection and conservation services for more than 50 museums and cultural institutions.

库房

中国丝绸博物馆的库房共分三层，其中可以分为一二级品库、历史文物库、低温库、时装库和机具库等。库房东侧还设有藏品管理室，特别要说明的是其中还设有研修室、文物观摩室，专门为前来从库房提取藏品进行鉴赏或研究的学者和爱好者们提供服务。

Storage

The storage of China National Silk Museum has three floors, providing space for the storage of first grade and second grade collection, the historical relics storage, the low-temperature storage, the fashion storage and the equipment storage. There is also a collection management room on the east side of the storage. It should be specially noted that there is also a study room (also called seminar room), which provides service for scholars and interested in who come for the appreciation or research of the collections.

八 园林 Garden

蚕乡桑庐

桑庐位于园区北侧，原型来自杭嘉湖平原的海宁云龙村，果树来自良渚大观山，并由中国丝绸博物馆与云龙村合作举办各种蚕事活动，如祭祀蚕神、唱蚕花戏、裹蚕讯粽、演皮影戏等活动处处体现了乡风民俗，做蚕具、结蚕网、缫土丝、翻丝绵等劳作体现了传统技艺，人们可以在这里领略和体验蚕乡风俗。

Sericulture House

The Sericulture House is located on the north side of the museum. The prototype is from Haining Yunlong Village in the Hangjiahu Plain, and the fruit trees come from Liangzhu Grand View Mountain.China National Silk Museum cooperates with Yunlong Village to hold various kinds of silkworm activities, such as worshipping silkworm gods, singing silkworm songs, making rice dumplings, performing the shadow play, and other activities that embody the folk customs of the township. Making silkworm rearing instruments, knitting silkworm nets, reeling mulberry silk and other labors reflect the traditional skills, and people can enjoy and experience the customs of silkworm villages here.

嫘祖像

嫘祖是传说中黄帝的元妃，中国蚕桑丝绸的发明者和传播者。此像由杨伯祥等于1992年设计制作，位于中国丝绸博物馆正门广场。2015年广场下挖地下车库，特移于此。

Lei Zu Statue

Lei Zu, the legendary Yellow Emperor's queen, is the inventor and communicator of sericulture and silk in China. This sculpture, once located on the main square of China National Silk Museum, was designed by Yang Boxiang et al. in 1992. It has been moved to the current place as the underground park was dug in 2015.

朱新予像

朱新予（1906–1986）浙江萧山人，是中国丝绸实业家和教育家，也是中国丝绸博物馆的倡建者。

Zhu Xinyu Statue

Zhu Xinyu (1906-1986), a native of Xiaoshan, Zhejiang, is a Chinese silk industrialist and educator, one of the advocators, and also for founding the China National Silk Museum.

染草园

染草园建立于 2016 年夏天，目前已经试种了 10 种常见的染料植物。春夏之交，在这个染草园可以收获红花、栀子、菘蓝、凤仙花和槐米。我们计划在 2020 年底能够成功种植 30 种染料植物，其中包括从国外引种的染草。

在染草园种植的植物一部分用作中国丝绸博物馆开展培训体验课的染材，另一小部分将用作科学研究的参考标本。此外，我们为游客，特别是年轻人提供学习辨识染料植物并参加染色体验课的机会，使大家能够明白天然染色是当今社会一种环境可持续发展的生活方式。

Dye Garden

The natural dye garden was constructed in the summer of 2016, in which about 10 common dye plants have been tentatively grown so far. In the small garden, safflower, gardenia, woad, henna and buds of pagoda tree can be harvested during the period of spring and summer. We plan to grow more dye plants including several species from foreign countries in the near future, and expect that 30 dye plants will successfully flower and fruit at the end of 2020.

The dye plants grown in the garden will be partially used to dye yarns and threads for some workshops hosted by China National Silk Museum every year, and a small part of plants will be collected and employed for references in scientific researches. Moreover, we provide an opportunity to learn how to identify natural dye plants in the garden and then participate in dyeing workshops for tourists, in particular young people who will understand dyeing with natural products is a way that is environmentally sustainable in the modern society.

图书在版编目(CIP)数据

中国丝绸博物馆:汉英对照 / 中国丝绸博物馆编.
—杭州:浙江大学出版社,2018.11
ISBN 978-7-308-18214-0

Ⅰ.①中… Ⅱ.①中… Ⅲ.①丝绸—博物馆—介绍—中国—汉、英 Ⅳ.① TS146-282.551

中国版本图书馆CIP数据核字(2018)第095046号

中国丝绸博物馆
中国丝绸博物馆 编

责任编辑	包灵灵
责任校对	黄静芬
封面设计	赵 帆
出版发行	浙江大学出版社
	(杭州市天目山路148号 邮政编码310007)
	(网址:http://www.zjupress.com)
排 版	城色设计
印 刷	浙江海虹彩色印务有限公司
开 本	787mm×960mm 1/16
印 张	4.75
字 数	50千
版 印 次	2018年11月第1版 2018年11月第1次印刷
书 号	ISBN 978-7-308-18214-0
定 价	50.00元

审图号:GS(2018)5193号

版权所有 翻印必究 印装差错 负责调换
浙江大学出版社发行中心联系方式:0571-88925591;http://zjdxcbs.tmall.com